DEPARTMENT OF NATIONAL RESOURCES
BUREAU OF MINERAL RESOURCES, GEOLOGY AND GEOPHYSICS

REPORT 173

Recharge conditions and the siting of bores in fractured-rock aquifers of the A.C.T.

G. M. BURTON

AUSTRALIAN GOVERNMENT PUBLISHING SERVICE
CANBERRA, 1977

DEPARTMENT OF NATIONAL RESOURCES

MINISTER: THE RT HON. J. D. ANTHONY, M.P.

SECRETARY: J. SCULLY

BUREAU OF MINERAL RESOURCES, GEOLOGY AND GEOPHYSICS

DIRECTOR: L. C. NOAKES, O.B.E.

ASSISTANT DIRECTOR, GEOLOGICAL BRANCH: J. N. CASEY

Published for the Bureau of Mineral Resources, Geology and Geophysics by the Australian Government Publishing Service

ISBN 0 642 02900 8

MANUSCRIPT RECEIVED: JULY 1973

REVISED MANUSCRIPT RECEIVED: JUNE 1975

ISSUED: OCTOBER 1977

Printed by Graphic Services Pty Ltd. 516-518 Grand Junction Road, Northfield. S.A. 5085

FOREWORD

G.M. Burton had proposed to make some amendments to the report that follows to bring it up to date; but he died in November 1972, and rather than amend his work the Bureau has decided to issue the report as a memorial, however inadequate, to an esteemed colleague and a noted hydrogeologist.

L.C. NOAKES
Director.

CONTENTS

TABLES

FIGURES

SUMMARY

Bores are commonly used to supply part of the water requirements of farms in the Australian Capital Territory (A.C.T.) and the adjoining areas of New South Wales. Most of the groundwater comes from fractured and weathered zones in crystalline rock.

Geologists of the Bureau of Mineral Resources site bores for local graziers and investigate groundwater drainage problems. In the course of this work they have systematically investigated the occurrence of groundwater and the natural environment which controls it. They have paid particular attention to the genesis of the crystalline-rock aquifers and the nature of their recharge.

This paper sets out some of the information gathered so far during this work.

INTRODUCTION

Since 1954 geologists of the Engineering Geology Group of the Bureau of Mineral Resources have investigated intermittently the occurrence of groundwater in the Australian Capital Territory. The inland part of the Territory covers 800 square miles of the tableland and alps of south-eastern Australia. The smaller coastal area of the Commonwealth Territory at Jervis Bay, which is in a completely different geological setting, is not covered in this paper. The work falls into three categories:

1. The location of groundwater supplies for farms and small settlements in the A.C.T., and in adjoining parts of N.S.W. by agreement with the New South Wales Geological Survey;
2. Studies of underground water as it affects engineering structures or gives rise to drainage problems; and
3. Research into aspects of the occurrence of water in crystalline rock.

The work has required a detailed study not only of the geology of the aquifers but also of all of the other elements of the local hydrological cycle (Fig. 1).

The development of supplies on farms is the most important activity and has received the greatest attention. Nearly all the bores for farm water supplies tap aquifers in fractured crystalline rock.

Crystalline rock aquifers and their development

A crystalline rock is defined for the purpose of this paper as one whose component grains have crystallized, recrystallized, or compacted to give a dense fabric which possesses no significant intergranular porosity. Included under this definition are plutonic, hypabyssal, and volcanic igneous rocks, marmorized limestones and metamorphosed or strongly re-cemented sediments. The character of the aquifers in such rocks depends on the porosity and permeability of joints and fractures, which are commonly modified by solution and weathering. Such aquifers are generally less regular in extent and permeability than the better known aquifers provided by alluvial deposits and sedimentary rocks.

Crystalline rock aquifers are now receiving more attention for farm supplies in Australia. In the past, rural landholders generally settled near perennial streams and springs, which together with roof tanks provided sufficient water for their domestic, pastoral, and agricultural needs.

Closer settlement in many areas has fully developed most of the land with natural perennial surface supplies. At the same time domestic consumption of water on farms has risen with the introduction of septic tanks, hot water systems, washing machines, evaporative-type air-coolers, gardens and, in some cases, swimming pools. Improvements in pastures and animal husbandry also require more, and better spaced, watering points; changes in farm economics have also led to diversifications which require additional water.

Figure 2 shows the main areas of crystalline rock in Australia and their relation to the main elements of climate. It will be seen that the main elements of climate vary markedly over a distance of a hundred miles or so. When these changes of climate are linked with changes in lithology and of tectonic environment, considerable variation in hydrogeology can be expected. The findings set out in this paper should not be extrapolated to other areas without due care; for example, in the A.C.T., changes of evaporation resulting from variation in elevation have been found to produce considerable changes in the recharge cycle between points as close as 30 miles apart.

Figure 2 also shows that the rainfall is both sufficient and reliable enough to provide good supplies of surface water in roof tanks and well sited earth dams in the A.C.T.; however, bores are usually needed to fully develop a farm and make it more drought resistant. The time at which the bores should be sunk depends very much on the relative merits of each water source, and the hydrogeologist has to be aware of all relevant factors (see Table 2) so that he can advise a farmer whether or not to defer sinking a bore, particularly if he considers that a deep and expensive bore, or a bore with a low yield or poor quality water, is likely to result.

The chance of failure in drilling for water in the A.C.T. and environs is low (5%) as can be seen from Table 1. Careful siting, however, can help considerably in finding the cheapest and best bore site. Between 1961 and 1967 the number of bores in igneous rock doubled, but the average depth for these bores from 1944 to 1961 and from 1944 to 1967 has fallen from 95 feet to 82 feet respectively; it is believed that part at least of this improvement is due to local hydrogeological research.

TABLE 1. WATER BORES IN A.C.T. AND ENVIRONS
(TO MAY 1967)

Country Rock	Number of Bores	Salinity Range (ppm)	Successful Bores		Unsuccessful Bores	
			Number	Average Depth (ft)	Number	Average Depth (ft)
Crystalline igneous	46	290-2240	44	82 ft	2	69 ft
Crystalline sedimentary	14	624-1560	13	199 ft	1	345 ft
Limestone	2	300-700	2	115 ft	-	-

TABLE 2. RELATIVE MERITS OF A.C.T. FARM WATER SUPPLIES
(Costs approximate only)

	Earth-dam		Roof-tank		Bore		Well	
Usual size	1500-2000 cub. yds. (250 000 - 350 000 gals.) on 5-acre catchment		8000 gals.		100 ft deep		10 ft deep	
	1967	1972	1967	1972	1967	1972	1967	1972
Cost: (1) Supply point	$400	$600	$540	$600	$900	$1100	$220	$300
(2) Pump, motor and housing	$300	$400	-		$700	$900	$300	$400
(3) Low tension electric supply per 100 yds.	$200	$250	-		$200	$250	$200	$250
(4) Polythene piping to house etc. per foot.	15c	15c	-		15c	15c	13c	15c
(5) Total cost of equipment supplying house 200 yds. from bore etc. and 100 yds. from power supply	$990	$1340	$540	$600	$1890	$2340	$800	$1130
(6) Annual depreciation and maintenance.	$80	$107	$35	$44	$100	$125	$20	$28
(7) Pumping per 100 gals.								
(a) to surface	(a) -		-	-	0.5 - 1.0 c (?)		-	
(b) to point further 100 ft. above.	(b) 0.5 - 1.0 c (?)		-	-	0.5 - 1.0 c (?)		0.5 - 1.0 c (?)	
Loss of water by:								
(1) evaporation	High		Very low		Nil		Nil	
(2) drainage	Low to very high		Nil		Nil		Nil	
Effect of drought	Very serious		Serious		Noticeable but not serious		Noticeable, may be serious in very bad drought.	

TABLE 2 (Continued)

Supply per annum/o	100 000 gals. ? (nett after evaporation)*	20 000 gals.	300 000 gals.	50 000 gals.
Contamination:				
(1) Suspended matter	Strongly discoloured	Clear	Clear	Clear
(2) Animal pollution	Great	Slight to great by rodents and birds	Slight	Slight
(3) Human	Nil to great	Nil	Nil to great	Nil to great
(4) Industrial	Nil to great	Nil	Nil to great	Nil to great
(5) Mineral	Medium	Nil	Medium to great	Medium
Suitable for use by:	Stock, irrigation, septic tank	All purposes	Often all purposes; each bore should be tested before use. Special attention should be paid to the analysis before using in expensive machinery such as washing machines and hot water systems.	Generally all purposes; each well should be tested before use.

* Yield depends markedly on slope and the type of soil and pasture in the catchment.

/o Average consumption per house: (1) Yass: 90 000 gals. per annum (consumption, 1959)
Canberra: 135 000 gals. per annum (design rate for town water supply, 1966)

NATURAL ENVIRONMENT OF GROUNDWATER

Geology

The Australian Capital Territory lies in a belt of Lower Palaeozoic sedimentary and volcanic rocks with a northerly trend. The oldest known rocks are Ordovician, but the presence of Cambrian rocks is suspected. The description of the geology of the A.C.T. (Fig. 3) is based mainly on the work of Opik (1954, 1958) and Noakes (1954).

The oldest sediments consist mainly of deep-water fine-grained slope and trough greywacke, slate, and claystone which were strongly folded and faulted during the Benambran Orogeny at about the close of the Ordovician. During the orogeny a geanticline formed in the vicinity of Canberra resulted in a local change from trough to shelf sedimentation, and as a result the Silurian Period is represented in the A.C.T. by shale, sandstone, limestone, and volcanic rocks. There was a gradual increase in volcanicity during the Silurian Period. Some tuffs are present in the Lower Silurian, and tuffs and flows are more abundant in the Middle Silurian, while the Upper Silurian contains numerous flows, tuffs, and intrusives.

The Bowning Orogeny strongly fractured the Silurian and Ordovician rocks which lay at fairly shallow depths (much less than 10 000 feet), and were hence more subject to the processes of fracture than to flow. It also consolidated the Canberra Welt (Noakes, 1954) which underwent little tectonic dislocation after the close of the Silurian.

The Devonian is mainly represented by volcanic rocks which, petrologically, are similar to the Silurian. There is, however, a marked difference in the degree of deformation of the rocks of the two ages in the area east of the Murrumbidgee River. The Silurian rocks are strongly fractured whereas those of Devonian age are relatively unfractured because they were protected from the effects of the Tabberaberan Orogeny by the stability of the underlying welt. West of the Murrumbidgee, Devonian rocks were not on the welt and Silurian and Devonian rocks differ little in structural deformation.

The only post-Devonian strata in the Territory are local thin gravel deposits, Cainozoic alluvium and widespread soils, some of which are 20 feet thick.

Topography and physiography

The central topographic feature of the A.C.T. is the northerly-flowing, incised Murrumbidgee River. Into it from the east flow the Molonglo River and its tributaries, the Queanbeyan River and Jerrabomberra Creek, which drain the hill and plain country (elevation 1800 to 2700 feet) around Canberra and the Gourock Highlands (2300 to 5300 feet) to the east and south of Canberra. The country to the west of the Murrumbidgee is rugged and mountainous (2000 to 6000 feet); it is drained by the Cotter, Paddys, Gudgenby and Naas Rivers. The topography and drainage are illustrated in Figure 4.

The physiographic evolution of the region has been discussed frequently in geological literature, but so far no completely satisfactory account has been published.

Certain broad general features which are useful for the hydrogeologist can be noted, however;

1. Much of the ridge and plain country around Canberra (in what Opik (1958) calls the Canberra Rift) is essentially an old mature land surface that had reached an advanced stage of maturity even in Lower Devonian time. Long periods of slow weathering and gentle erosion reduced the area to a peneplain in late Tertiary time. The surface is rejuvenated in the area adjoining the Murrumbidgee River. The ridge and plain country was subject to considerable periods of wasting and soil formation in Pleistocene and Recent time.

2. The mountains west of the Murrumbidgee are comparatively youthful but possess puzzling areas of more mature topography, notably the valleys of the Gudgenby and Paddys Rivers and the Uriarra area.

3. The highlands east of Canberra are mature but possess thinner superficial deposits and show more rejuvenation than the Canberra plains.

4. The highlands south of Canberra appear to be a unit containing many of the mature features of the Canberra Plain and some of the more youthful features of the eastern highlands.

5. The rejuvenation which is most noticeable near the Murrumbidgee River, probably occurred since late Pliocene time and initiated the upstream migration of nickpoints along the Murrumbidgee and its tributaries. Some local meridional and northeasterly trending faults also developed during the same period and superimposed their own pattern of rejuvenation and nickpoints.

6. The nickpoints developed into features, similar to mexican dams, which the writer calls 'nickpoint bars'. Zones of fresh rock beneath the previous *potentiometric surface were exposed at the nickpoint. Soil subsequently developed over many of the nickpoints, particularly those in the higher reaches of the catchments. The soils were commonly pedocalcic types of low permeability and enhanced the groundwater barrier at the nickpoint. The nickpoint bars maintained high groundwater level in the basins above the bars; because these upper basins were commonly surrounded by low hills with thin skeletal soils conducive to high infiltration, mass-wasting rather than erosion occurred in the basins. Erosion, however, predominated at times and partly stripped the basin before a new period of mass-wasting mantled the area again.

A better knowledge of the evolution of the physiography would be most useful. The hydrogeologist working on crystalline rocks is primarily concerned with the upper 300 feet of the Earth's crust and is interested in the nature and duration of the processes which have not only formed the topography, but also considerably modified the underlying rocks and their ability to store and convey water.

It is important for the hydrogeologist to recognize the pattern of fossil potentiometric surfaces defining old lower surfaces of weathering, and to establish, at least partly, the subsequent episodes of weathering, mass-wasting, deflation and erosion that have prevailed in different segments of catchments. Continuing advances in our knowledge of the influence of mineralogy and geographic aspect on weathering are helping in this work.

The task of unravelling the picture however, is difficult in the complex geology and geomorphology of the A.C.T. The writer believes that important future advances will come from accurate mapping of the relatively well known system of meridional faults and the less well known movements in both systems. The most recent fault movements have largely determined the rate and extent of erosion, and the deposition of superficial deposits, near each fault. The faults changed erosion and deposition rates by truncating some drainage basins and inducing piracy of others. The formation of fault scarps has not only changed river grades, but has also affected the areal distribution of rainfall.

* <u>Potentiometric Surface</u>. The potentiometric surface, which replaces the term 'piezometric surface', is a surface with represents static head. As related to an aquifer, it is defined by the levels to which water will rise in tightly cased wells.

These points emphasize the fact that the essential basis for hydrogeology is detailed geology and geomorphology. Even though the second edition of the Canberra 1:250 000 Sheet was published in 1965 only 30 percent of the mapping has been classed as detailed reconnaissance or better; this is understandable in view of the complexity of structure, the lithological similarity in much of the poorly fossiliferous sediments, and the chemical similarity of many of the igneous rocks.

Climate

The Australian Capital Territory lies in a meteorological zone in which the rainfall is both reliable and evenly distributed throughout the year (Figs. 2 & 5). Most of the Territory receives an average rainfall of between 18 and 25 inches per year. The mountainous areas in the west, however, receive as much as 60 inches of precipitation (including snow) per year. Most bores lie in an area having an average of about 25 inches of rain per year; evaporation in the same area averages 51 inches per year.

The most significant feature of the climate is the annual range of evaporation. Whereas the rainfall is evenly distributed, evaporation rises sharply in November and declines equally sharply at the end of April. The beginning of the winter season in May with the onset of frosts and the end of the growing season for grasses coincides with the sharp drop in transpiration and evaporation, and soil moisture rapidly increases; transpiration of trees, however, continues and a different soil moisture regime prevails in forest catchments. Consequently, the year can be divided into two main hydrological seasons of six months each according to soil moisture, the 'winter' or wet season beginning in May, and the 'summer' or dry beginning in November.

Water Provinces

The A.C.T. is subdivided on the basis of differences in geology and physiography into six water provinces each having significant individual differences in groundwater regime. The six provinces are shown in Figures 3 & 4 and the main characteristics of each province are set out in Table 3.

TABLE 3. A.C.T. UNDERGROUND WATER PROVINCES

Province	Topography	Principal Rock Types	Approx. Annual Precipitation (mm)	Aquifers* Good	Aquifers* Doubtful	Aquicludes*	Aquifuges*	Depth to Potentiometric Surface (m)	General Salinity Range (ppm)
East Gourock Highlands	Mature valleys noticeably rejuvenated in lower reaches; well defined rounded ridges	Strongly folded quartz greywacke, siltstone, and slate	25-27	Greywacke and siltstone	Slate, porous soils	Clay soils, weathered slate	Probably slate in nonfaulted areas below about 100 m	5-70'?	700-2000
West Gourock Highlands	Mature valleys noticeably rejuvenated in lower reaches; well defined rounded ridges	Volcanic and intrusive porphyry, shale and limestone	18-24	Stressed porphyry	Unstressed porphyry, granite, limestone, porous soils	Clay soils	"	5-50'	300-2000
Canberra Ridges and Plains	Mature undulating plain country with ridges of more resistant rock. Shows marked rejuvenation near the Murrumbidgee River	Sediments, volcanics, and intrusive porphyry	24	Stressed porphyry, volcanics, and competent sedimentary rocks; coarse alluvium	Unstressed volcanics, porphyry, granite, porous soils and alluvium	Clay soils, alluvial clay	"	Usually 0-50' Some areas of sediments with poor recharge 50-90'	300-2000
Murrumbidgee Scarp	Isolated ridge with moderately deep dissection	Hornfels, siltstone, greywacke, slate, and granite	18-30	-	Hornfels, greywacke, stressed granite, and pockets of scree	-	Unstressed granite?	20-100'	?
Paddys River and Gudgenby	Rounded ridges and moderately broad valleys. Rejuvenated near Murrumbidgee River	Gudgenby; granite. Paddys River: granite and volcanics	30-35	Stressed or closely jointed granite and volcanics	Porous soils and weathered granite and volcanics	Clay soils	Unstressed granite	0-100'	300-2000?
Bimberi Mountain	Rugged mountainous country	Granite and highly folded greywacke, siltstone, and slate	35-60	-	Greywacke, siltstone, stressed granite, volcanics, permeable soils, and scree pockets	-	Unstressed igneous rocks?	0-150'?	300-2000?

* Aquifer: a geological formation which is capable of transmitting appreciable quantities of water under normal field conditions.

* Aquiclude: a geological formation which may contain groundwater but is incapable of transmitting significant quantities of it.

* Aquifuge: a geological formation which neither contains nor transmits significant quantities of groundwater.

DEVELOPMENT OF GROUNDWATER RESOURCES

In developing the groundwater resources of the A.C.T. the hydrogeologist is interested particularly in:

(i) the ability of the aquifer to store and transmit water

(ii) the water balance of the aquifer

(iii) the quality of the groundwater

(iv) the economic exploitation of the aquifer.

The Aquifer

For the purpose of this paper discussions on aquifers will be restricted to the crystalline or fractured rocks; the important, but less widely distributed alluvial and lacustrine aquifers of Lake George and the Molonglo Valley will not be treated.

The ability of the fractured rocks to store and transmit water depends mainly on the permeability and porosity of fractured zones, weathered zones and solution cavities; some sediments and tuffs also have minor residual intergranular porosity.

A knowledge of the progressive development of the permeability and porosity of the fractured rocks is important in that it enables the geologist to select the most important target to drill for any bore. The picture as determined so far by the writer in the A.C.T. is summarized in Table 4. The table is quite detailed but still somewhat generalized.

The relative yields of different types of fracture is next in importance; the approximate ranges of specific capacities (2 hours after pumping commences) of bores in some types so far determined are shown in Table 5.

TABLE 4. PHASES IN FORMATION OF FRACTURE PERMEABILITY IN THE A.C.T.

ROCK TYPE	PHASE 1	PHASE 2		PHASE 3	PHASE 4	
	DIAGENETIC ETC.	OROGENIC		EPEIROGENIC	WEATHERING	
		EARLY	LATE*		PHYSICAL	CHEMICAL
SEDIMENTS	Negligible fractures. A few due to differential compaction. Formation of or filling of solution cavities. Good intergranular porosity and permeability retained	Some joints and faults formed. Variability of competence of adjacent beds produces greater jointing. Some intergranular porosity retained	Many major and minor tensional and compressional faults and joints and shear zones formed; tensional, rotational and shear joints formed in folded competent beds. Most of remaining intergranular porosity destroyed	Increased permeability of most existing joints and faults; new local fracture zones developed	Unloading during erosion opens existing joints and creates many new ones, particularly when unloading process is coupled with expansion of mantle due to permafrost and weathering. Stresses due to insolation and tides may contribute significantly	Permeability destroyed but increased porosity in upper strongly weathered zones of susceptible rocks, such as shale. Weathering may produce jointing of less susceptible but competent rock (e.g. sandstone) by partial expansion of rock itself or adjacent susceptible rocks
VOLCANICS INTERBEDDED WITH, OR OVERLYING, LESS COMPETENT SEDIMENTS	Some bedding and cooling joints formed. Formation of or filling of solution cavities in sediments. Good intergranular porosity and permeability retained in some sediments and tuffs. Vesicular porosity and permeability retained in some volcanics	Some cooling and bedding joints partly destroyed by welding. Some fracture cleavage developed, other joints developed, particularly in sequence of beds of different competence	As above, but size of existing cooling and bedding joints may be further reduced	As above	As above	As above. Sheeting joints with good permeability may be developed at transitional zone between weathered mantle and fresh rocks in massive volcanics
VOLCANICS	Some bedding and cooling joints formed. Fair intergranular porosity and permeability retained in some tuffs. Vesicular porosity and permeability retained in some volcanics	Intergranular and vesicular permeability of tuffs and lavas partly destroyed. Some bedding and cooling joints also partly destroyed. Some fracture cleavage and other joints developed in sequences of volcanics of different competence	Numerous major and minor tensional and compressional faults and shear zones formed; fracture cleavage and tensional, rotational, and shear joints formed in some sequences. Some cooling and bedding joints partly destroyed Most of remaining intergranular and vesicular permeability destroyed	As above	As above	Permeability destroyed but porosity in strongly weathered zones of volcanics increased. Permeability and porosity in most existing fractures in partly weathered zones improved. Sheeting joints with good permeability may develop at transitional zone between weathered mantle and fresh rock in massive volcanics

TABLE 4 (Continued)

PLUTONIC AND HYPABYSSAL IGNEOUS	Some contraction joints formed near margins; may be partly filled by veins. Zones of autometamorphism may have appreciable porosity	Generally emplaced late in orogeny and hence not affected by early orogenic events	Partly affected by last events of orogeny, such as diapiric intrusion of solidified igneous rock	Important stage in development of local fracture zones and increased number and size of joints in scarce existing fracture zones	Tectonic joints are scarce, but during weathering widespread joint zones are formed in plutonic and hypabyssal rocks	
					A relatively small number of joints with good permeability are formed by stresses due to unloading, coupled with expansion of mantle due to weathering, perma-frost, insolation and tides.	Zones of good porosity but poor permeability formed in strongly weathered rocks. Permeability and porosity of stress joints, previously formed by physical weathering in partly weathered zones, improved. Sheeting joints with good permeability may develop at transitional zone between weathered mantle and fresh rock

* Pneumatolitic injections in late stages of orogeny may fill and partly seal many joints very close to major igneous intrusions.

TABLE 5. SPECIFIC CAPACITIES OF BORES IN THE A.C.T. AND ENVIRONS

Lithology	Type of Fracture Yielding Supply	Approximate Specific Capacity of Bore (g.p.h. per ft) (drawdown at 2 hours)
Igneous rock	Major sheeting and rebound joint zones to depth of about 30 m below surface	15 - 40
	Minor fracture zones to about 30 m below surface	10 - 20
	Minor cooling joints and individual tectonic fractures from 30-75 m below surface	2 - 10
	Major fracture zones from 6-90 m below surface	15 - 50
Limestone	Well developed solution cavities (but not caves)	50 - 400?
Interbedded siltstone and slate	Minor fracture zones and individual tectonic fractures, 5 to 300 feet below surface	2 - 10
	Medium to major fracture zone, 5 to 300 feet below surface	10 - 20

The yield of bores in the A.C.T. after 3 hours pumping is generally between 100 and 1800 gallons per hour. One recent experimental bore drilled by the Bureau at its Fyshwick Depot, however, yielded more than 9000 gallons per hour from a major fracture zone about 260 feet below the potentiometric surface; this is the only local bore known to exceed 2000 g.p.h.

The general pumping characteristics of bores depend not only on local permeability but also on the presence nearby of major zones of weathering, which provide 'leaky' aquifer conditions. Some typical examples of tests carried out by Wilson (1960) for the Bureau are shown in Figure 6. Some of these bores were developed during, and by, the tests, which were aimed at determining the decline in yield under maximum drawdown. It was not always possible to maintain maximum drawdown throughout the full tests.

Longer tests carried out by the writer at varying pumping rates achieving near maximum drawdown at the Hall School Bore are shown on Figure 7. This bore is a typical medium standard bore which is pumped heavily for much of the year. A meter was fitted to the bore on completion and the yield recorded for several years. During the drought year of 1965 a total of 490 000 gallons was pumped; the weekly consumption commonly exceeded 20 000 gallons and reached a peak of 34 000 gallons. Meters have been fitted to several other bores in the region in order to gather similar useful statistics.

Water balance

The aquifer must not only store and transmit water, but it must have a suitable water balance: it must have suitable recharge and must not have excessive natural or artificial discharge. The water balance equation, with emphasis on the more important elements in the A.C.T., is set out below. The unit area of application for the equations is a region for which lateral losses and gains are minor and are not shown.

1. (General equation)

Change in groundwater storage = (Groundwater recharge) - (Groundwater loss)

2. (Detailed equation - the most important parameters are shown in capitals, and the items next in importance are underlined)

CHANGE IN GROUNDWATER STORAGE = PRECIPITATION

- (minus)
  - (a) Run-off and overland storage
  - (b) Increment to vadose zone
  - (c) EVAPO-TRANSPIRATION

- (minus)
  - (a) GROUNDWATER DISCHARGE to
    - (1) EFFLUENT STREAMS
    - (2) Springs
  - (b) Groundwater loss to vadose zone by:
    - (1) Capillary action
    - (2) Vapour movement
    - (3) Biological consumption
  - (c) Discharge by bores and wells

This equation may be expressed algebraically, with symbols for the corresponding terms, as follows:

$$GS = (P - (SW + VW + ET)) - ((DE + DS) + (LC + LV + LB) + DB)$$
$$= P - (SW + VW + ET) + DE + DS + LC + LV + LB + DB)$$

In the study of the water balance of an aquifer, it is the trend in groundwater storage that is important, and this is measurable as fluctuations in the potentiometric surface. Precipitation and evaporation potential are directly measurable, and stream gauging provides a check on run-off and groundwater discharge to effluent streams. An examination of such observations provides a better appreciation of aquifer potential, and gives an indication of the bore specifications for tapping the aquifer.

The Bureau sank several bores in late 1958 (Fig. 8) to study seasonal changes in the potentiometric surface. Some of the observations from two of the bores are shown in Figure 9. In view of the considerable amplitude of the changes and the variation between bores, it was decided to expand gradually the network, using the Bureau's own drills in slack periods together with observations on some privately-owned bores. The network was systematically expanded to catchments with different climates, geology and vegetation, to isolate the effects of the several parameters. The more important details of the network are set out in Table 6.

Geologists are now able to predict quite accurately the amplitude of fluctuation of the potentiometric surface at most bore sites, to define a suitable depth for the bore and to assess with greater confidence its safe yield.

Other important deductions to date from the network are:

1. The main period of aquifer recharge in the average year extends from mid-June to about November on the Plains (elevation about 2000 feet), and is several months longer in the Mountains (at 4000 feet).

2. The aquifer recharge in forest catchments on the plains is less than in the grassed catchments.

3. Geomorphology, as much as geology, controls not only aquifer recharge, but also discharge.

Several other trends are also under observation.

TABLE 6. REGIONAL GROUNDWATER OBSERVATION BORES A.C.T. AND ENVIRONS

Bore & Location	Groundwater Province (Figs 3 & 4)	Elevation of Bore (ft)	Nature of Catchment				Gauging Started	Depth to Water (1/2/67) (ft)	Maximum Change in Levels (ft)	Approx. Yield (g.p.h.)
			Ann. Rainfall (in inches)	Geology	Physiography	Vegetation & Culture				
Belconnen 5 (Gribbs Farm)	Ridges & plains	2140	25	Intrusive Porphyry	Higher slopes of low divide	Natural grassland	Dec. 1958	48.8 ft	22 ft	200
Belconnen 6 7 & 8 (CSIRO Farm)	" "	1940	25	Intrusive Porphyry	Small perched basin with granite bar	Grasslands, partly pasture improved	Dec. 1958	9.3 (Bore 8)	10	1500 (6) 20 (7) 900 (8)
Jeir 1 (Jeir Station)	" "	1875	25	Acid volcanics, medium dip	Margin of broad plain	Mainly natural grassland	Oct. 1961	19.7	5	100
Belconnen 13 (Black Mountain)	" "	2032.6	25	Siltstone & slate, strongly folded	Lower slopes of strong dividing ridge	Mainly natural eucalyptus forest	March 1966	88.0	1	20
City 13 (BMR Fyshwick)	" "	1895	25	Volcanics & sediments, strongly sheared	Low ridge on rolling plain	Natural grassland, partly developed as light industrial area	June 1966	19.7	6	10 000
Lanyon 5 ('Melrose Valley')	W. Gourock Highland	2350	25	Porphyry	Lower slopes of major perched valley, partly rejuvenated	Grassland, pasture improved with clover	Aug. 1960	36.9	13	100
Tennent 1 (Honeysuckle Tracking Station)	Gudgenby	3520.8	35	Granite, deeply weathered on possible lineament	Lower slopes of deeply dissected valley	Mainly natural eucalyptus forest	April 1966	22.5	2	100
Cotter River 1 (Corin Dam road)	Mt. Bimberi & Paddys River	4064.2	40	Granite, deeply weathered on possible lineament	Major saddle in major dividing range	Natural eucalyptus forest	April 1966	7.2	14	100

In addition to the regional network of observation bores, the Bureau maintains 8 observation bores around Lake Burley Griffin. Seven of the bores were commissioned by the National Capital Development Commission, acting on the advice of the Bureau; they were established as the lake filled. The eighth was drilled by the Bureau at its head-office as an observation bore and instrument testing bore.

All data from both networks of observation bores, together with data on rainfall, evaporation, air temperature, soil temperatures and barometric pressures from the Yarralumla Climatological Station and data on water levels in Lake George and other river basins, are entered on punch cards. Graphs of the data are regularly up-dated on the Calcomp plotter of the C.S.I.R.O. computer. Log-plots are similarly prepared for some of the Lake observation bores.

Quality of the groundwater

It can be seen from the triangular diagrams of Figure 10 that the groundwater found in bores in the A.C.T. and environs is principally a calcium-magnesium-sodium or calcium-magnesium water, as far as the cations are concerned, and a bicarbonate-chloride or bicarbonate water, as far as the anions are concerned. The salinity of the water (Table 7, Fig. 10) is on the whole quite acceptable, with 53 percent of the analyses having a salt content of less than 800 parts per million (ppm) and only 5 percent of the analyses exceeding 1600 ppm. There are no known waters with dangerous proportions of fluorine or boron. Analyses of the groundwater and several surface waters are compared in Table 8.

TABLE 7. FREQUENCY DISTRIBUTION TOTAL DISSOLVES SOLIDS OF WATERS IN BORES IN A.C.T. AND ENVIRONS

Total Dissolved Solids (ppm)	Cumulative Number of Analyses	Cumulative (%)
400	12	25
800	25	53
1200	37	79
1600	45	95
1600	48	100

TABLE 8. WATER ANALYSES

Source	Total Dissolved Salts (ppm)	Conductivity 25°C (mhos/cm)	pH	Sodium Adsorption Ratio	Radical (Milliequivalents per litre)						
					Ca	Mg	Na	K	HCO3	SO4	Cl
Igneous rock, good quality water	290	450	7.3	1.0	1.3	0.9	1.1	-	2.7	-	0.8
Igneous rock, medium water	800	1200	7.0	1.5	7.1	4.8	3.6	-	8.6	1.0	5.1
Ordovician sediments, medium water	734	-	8.1	1.7	7.5	5.1	4.2	-	9.0	1.0	6.5
Ordovician sediments, poor water	1560	2100	7.0	2.3	9.6	9.1	7.4	-	10.3	6.4	9.0
Lake George, N.S.W.*	1286	2270	8.0	13.0	1.2	2.7	18.1	0.1	4.4	1.6	16.6
Lake Bathurst, N.S.W.	714	1310	9.3	6.9	0.8	2.5	8.7	0.5	2.8	0.4	8.8
Queanbeyan River,** Googong, N.S.W.	100-180	100-150	7-8	0.2	0.3	0.4	0.4	-	1.0	0.1	0.2

* Sample collected November, 1960, when maximum depth of lake was 13.5 feet.
** Inland stream draining sparsely populated area.

NOTE: The Molonglo River water due to pollution from mining activity is not a typical inland water, and its use for any purpose requires close examination.

The reasonably low concentration of salts and the bicarbonate nature of the water reflect the local hydrogeological conditions. The drainage network of effluent streams is quite dense and much of the groundwater moves only 1 to 2 miles from the recharge areas, in areas of thin skeletal soils on the hills, to its discharge as stream flow in the effluent streams nearby. Most of the movement is in well jointed rock less than 200 feet below the surface, and circulation is assisted by regular recharge.

The main disadvantage of the water for domestic or industrial use is the hardness - practically all waters fall within the classification of 'very hard' in the system used by the United States Geological Survey, as can be seen in Table 9; 51 percent of the samples exceed 600 ppm of hardness.

TABLE 9. FREQUENCY DISTRIBUTION OF HARDNESS OF WATER IN BORES OF A.C.T. AND ENVIRONS

Hardness (range: as ppm of $CaCO_3$)	Number of Analyses
Soft (0-60)	2
Moderately hard (61-120)	1
Hard (121-180)	1
Very hard (180)	43 (incl. 25 of hardness 600 p.p.m.)

When used with normal caution as to drainage, soil type and method of application, water from most bores is quite suitable for general agricultural purposes. Although the salt content is relatively high for irrigation purposes the rainfall and drainage are sufficient to prevent serious accumulation of salt on much of the agricultural land in the A.C.T. Further, the sodium adsorption ratio (S.A.R.)* is quite low (Table 10).

* Sodium-Adsorption Ratio (SAR). The SAR value for a water is simply related to the experimentally determined adsorption of sodium by soil to which the water is added. It is defined by the equation, $SAR = Na^+ / (Ca^{++} + Mg^{++})/2$, where Na^+, Ca^{++}, and Mg^{++} represent the concentrations in milli-equivalents per litre of the respective ions.

TABLE 10. FREQUENCY DISTRIBUTION OF SODIUM ADSORPTION RATIO (SAR) OF WATER IN BORES OF A.C.T. AND ENVIRONS

S.A.R.	Number of Analyses
2	37
2-4	10
4	0

The temperature of the groundwater is usually between 59 and 63°F.

Economic Exploitation

The economic exploitation of the aquifer depends on many hydrogeological and physical factors.

The problem usually facing the hydrogeologist is to find a bore site most suited to the farmer's requirements. This usually amounts to finding a site giving the best possible combination of:

1. suitable geology and geomorphology
2. easy, and minimum, drilling
3. closest to the point of consumption
4. largest supply
5. best water
6. greatest reliability in dry periods.

The economic and dependent hydrogeological factors involved in this are analysed in Table 11. If due regard is had for the hydrogeological factors in the table reliability of supply is generally assured.

TABLE 11. ECONOMIC FACTORS INFLUENCING SITES FOR BORES

Economic Factors of Construction and Operation	General Requirements		Consequent Hydrogeological Requirements
Low cost of construction	Easy drilling	Rapid penetration	Maximum percentage of self-supporting soil and weathered rock in profile and minimum thickness of very hard or abrasive rock.
		Minimum reaming	Avoidance of structures likely to deflect drilling.
	Minimum casing and screening	Minimum trouble from caving or swelling ground.	Avoidance of formations likely to cause caving, swelling etc.
	Minimum drilling	Minimum depth	High potentiometric surface Fractures of suitable permeability at optimum depth below potentiometric surface.
Low cost of pumping	Close to power source At optimum elevation Closest to point of consumption		Careful mapping outwards from point of consumption and power supply and from ground at optimum elevation.
Greatest return on capital outlay	Largest quantity		Thick zone of well developed, continuous fractures preferably near a zone of weathering with high specific yield and good permeability.
	Best quality	Best natural quality	Nearby major recharge and optimum discharge to prevent stagnation Path from recharge area free from harmful mineralization and infiltration by cyclic salt.
		Free from pollution	No pollution sources, such as septic tanks, disease-infected dams or agricultural chemical discharges, nearby. Sufficient soil cover etc. immediately near bore to purify or seal-off local pathogens.

SITING OF BORES

General

The siting of bores in the Canberra region depends on the knowledge of each of the elements set out under the Development of Groundwater Resources, above.

Cost and difficulty of boring are of great importance. The fresh crystalline-rock in zones where joints are few can be so hard that drilling is expensive and in some cases impracticable for percussion and some rotary drills: it is important for the geologist to select a site which will encounter the minimum section of the hardest rock. The introduction of the air-hammer drill, however, is giving the geologist greater latitude in the selection of economic sites. The air-hammer also permits him to drill deeper in hard rock; the increased available drawdown makes possible bores with larger yields.

Aerial photos are invaluable in siting bores in the Canberra region. The usual practice is to study the photos briefly with the large prismatic stereoscope before proceeding to the field. The photos are again used in the field with a pocket stereoscope to study the geology, geomorphology and hydrology of the area in which the site is required.

The usual procedure in siting is to consider the local recharge of the area, the geology and the topography, and then to estimate from previous experience the depth to the potentiometric surface (the potentiometric surface has a slope of about 1 in 30 to 1 in 40 over most of the plain country).

Allowance is made always for the estimated seasonal and long range fluctuation in the potentiometric surface. In this regard it is interesting to note the fluctuations of water levels in the bores in Figure 4. The area is also examined for geological features, such as faults, dykes and belts of shale, and geomorphological features, such as nickpoint bars, that would impede the discharge of groundwater and form areas of above-normal groundwater level.

A site is then selected wherein suitable joints or fractures will be encountered at least 20 to 50 feet below the lowest long range potentiometric surface and to have the maximum thickness above this of soil and weathered rock which is very easy to drill.

In limestone country, the slope of the potentiometric surface commonly is as low as 1 in 150 to 200, and great care is needed to avoid dry or very deep holes.

The procedure when searching for joints and fractures in the sedimentary rocks, such as the Ordovician, is to find first a suitable sequence containing sufficient competent rock which fractures easily, and then to locate the bore site in faulted or suitably folded and fractured areas of that rock; areas where hard steeply-dipping beds are likely to deflect the drill are avoided as much as possible.

In the fractured igneous rocks the problem, commonly, is to determine whether the exfoliation sheeting, and rebound joints, are well developed and lie sufficiently far below the potentiometric surface to give a good hydraulic head. If they do, they are the natural drilling target because they have good permeability and generally the most direct recharge; they carry the lowest salinity water (usually 200 to 400 p.p.m.).

If the shallow joints lie above the potentiometric surface it is necessary to proceed cautiously and to try to find an area which contains tectonic, or igneous contraction joints from 40 to 100 feet below the potentiometric surface. This usually requires careful field work, particularly if the area is underlain by little-fractured late-Silurian or Devonian intrusives, or Devonian volcanics.

Faulting must be considered carefully in Canberra where both high angle reverse faults and normal faults occur; faults may provide suitable drilling targets, but each should receive careful study. The reverse faults may be tight and may not yield water. It is better to site the bore in the adjoining subsidiary fractures if the fault has been determined as a reverse or compressional fault or if the nature of the fault is uncertain. One of the unsuccessful bores near the A.C.T. may be entirely within a tight compressional fault zone. Deep weathering also occurs in some faults and adequate permeability may not occur until a great depth is reached.

To assist Bureau geologists in the systematic selection of bore sites in the A.C.T. a short proforma Bore Advice Report (Appendix I) has recently been instituted. It is hoped that not only will this be of use in the siting of bores, but that it will help build up better hydrological data for the region.

Examples of simple sites

Figure 11 illustrates three bore sites under the simplest conditions in crystalline igneous rock, in hill and plain country around Canberra.

Site C is the least desirable site. It has poor local storage and poor recharge because much of the rainfall is carried away from the area by overlying, more permeable, surface joints and weathered rock (Zone 2). Drilling will commonly be relatively easy because of the depth of weathering. The bore will not encounter the potentiometric surface until it is in the zone of few tectonic or contraction joints. The potentiometric surface will be subject to considerable fluctuations (possibly as much as 30 feet) and the bore may fail in drought.

Site B is the most desirable site. It has the greatest thickness of soil, talus, lateritic products and decomposed rock (Zone 1). The potentiometric surface is shallow and at times artesian conditions occur. The bore will encounter the surface jointing and weathered permeable rock of Zone 2 at sufficient depth below the potentiometric surface. The best water supply is in the open joints of the slightly weathered rock at the base of this zone. The aquifer is subject to good recharge on the hillside, has a good hydraulic gradient and is subject to periodical flushing when surplus water escapes rapidly as springs in very high rainfall periods. It will not encounter much very hard fresh rock. Assessment of yield at this site should take into consideration possible fluctuations of as much as 10 feet (on present knowledge) in the potentiometric surface.

Site A is less desirable than B but more desirable than C. Zone 1 (soil and very decomposed rock) may be thin. Zone 2 (surface jointing and weathered rock) may have been partly removed by rejuvenation and the vertical joints may have been confined more in the absence of topographic relief. The bore will depend for its supply on water in the few tectonic and contraction joints of Zone 3. Water in the deeper sections of this zone will be more stagnant and have a higher salinity. The potentiometric surface is not likely to fluctuate much. The major creek is generally an effluent stream and unlikely to assist recharge or reduce salinity of most bores on the plain.

Zone 4 has not been discussed in regard to any of the sites. It is the zone, commonly deeper than 250 feet, in which joints are so few and tight that the change of obtaining water is usually low. Bores that reach this zone without encountering water should in most cases be abandoned and a fresh start made elsewhere.

Miscellaneous Geomorphological Examples

Figure 12 illustrates the relationship of some aspects of geomorphology to the suitability of bore sites in the A.C.T. The main recharge areas are on the thin skeletal soils of the ridges.

The further one proceeds down the catchments the greater the salinity becomes, but the smaller is the maximum seasonal and long range fluctuation in the potentiometric surface.

The valley containing Bores 2, 3 and 4 is a long narrow depression formed originally by rapid erosion along a fault; in subsequent periods of mass-wasting the valley slopes shed considerable detritus, but the surface run-off from the soil and the profile of the lower reaches of the valley were unsuitable for the removal of detritus; superficial deposits up to 20 feet thick will be found along the northeastern side of the creek.

Bores 2, 3 and 4 are located in small perched basins above minor nickpoint bars which differ from one another in the degree of physiographic restriction they impose. The restriction is least for Bore 4, but the restriction is reinforced by a permeability barrier at the fault and by the high general groundwater level in the lower reaches of the valley.

Bore 8 is situated at a major nickpoint bar where discharge is relatively slow, and is at a considerable distance from the main recharge areas on the skeletal soils higher up the valleys.

Bores 1, 5, 6, and 7 have small but nearby recharge areas; discharge is easy and rapid. The salinity of the water is low, but the seasonal and long range fluctuations of the potentiometric surface are great.

Miscellaneous Geological Examples

Figure 13 illustrates a group of miscellaneous examples where geology markedly influences the choice of site.

In the case of the weathered massive igneous rock (e.g. granite) much depends on the relationship of the potentiometric surface to the geology. If the potentiometric surface is shallow (I) Site 1 is preferable: the sheeting joints at the base of the weathered pocket are the natural target; Site 2 would require a considerably deeper bore in harder rock; its greater available drawdown would be needed to equal the greater local storage at Site 1 in the leaky weathered mantle. If the potentiometric surface is deeper a very much more carefully judged site would be needed to intercept the narrow fault well below the potentiometric surface.

In example 2, which has a major fault with a deep weathering zone along it, good permeability could be expected in the numerous fractures on both sides of the fault. The potentiometric surface on the recharge side would be high because of the damming effect of the deeply weathered fault zone. On the downstream side of the fault the incresed permeability and the possibly smaller flow of water would produce a deeper potentiometric surface; the quality of the water would also be poorer as sulphide mineralization is common near major faults in the A.C.T. (particularly those in the Middle Silurian volcanics) and deep circulation of water around the fault through such a zone can be detrimental. Site 5 is the best site: there is some soft rock in the section; permeability in numerous fractures is good and the potentiometric surface is high. Site 4 would be unsuitable because of poor permeability down to an excessive depth.

With bedded sediments the choice is more difficult. If the dips on the limbs of the folds are very steep the harder beds at Site 8 could deflect the drill off line and cause excessive delays for reaming, to straighten the hole. The sandier beds at Sites 7 and 9 would be preferable and much would depend on the geologist's judgment of available drawdown, recharge and discharge for each site.

Limestone sequences pose serious challenges. The potentiometric surface in such sequences (as distinct from isolated lenses) is usually deep. Solution cavities tend to develop along particular beds of limestone with suitable physical or chemical character, or along the lines of stress such as faults; once partly developed, these cavities are rapidly and preferentially enlarged. Site 11 on the axis of the syncline is preferable because it cuts a greater true thickness of limestone and has a higher chance per foot of drilling of cutting a suitable bed; it will also cut the cavities where a greater volume of water has passed and produced better cavities. It is not uncommon, however, to find that access to such a site is impossible because of the deep gullies that commonly follow the axis of the syncline. A bore at a site such as 12 would cut obliquely across the bedding thereby intersecting a smaller stratigraphic thickness per foot of hole than at Sites 10 or 11. Further, the drill would tend to be deflected and would possibly jam during drilling. Temporary mechanical failures could also occur in the finished bore due to movement of earth annd other debris in the steep cavities leading down to the bore casing: such failures have already been experienced.

DEVELOPMENT OF BORES

The development of bores in crystalline rocks by explosives, by acid (in limestones) and pumping, or by surging with Calgon all have their place, but the decision of when and what process to use requires considerable experience. The use of explosives for fracturing is the most common technique. The size and position of the charge should only be decided on after carefully inspecting the log of the hole.

ACKNOWLEDGEMENTS

Many graziers, but particularly the executors of the Gribble Estate, Dr R. Reader of 'Melrose Valley', and the proprietors of Jeir Station, have greatly assisted these groundwater investigations by willingly making bores on their properties available for hydrological measurements. The co-operation of Mr S. Ablamowicz of Pacific Boring Co., Yass, and Mr C. Nilon, Drilling Contractor, Queanbeyan, has been invaluable at all times.

REFERENCES

BURTON, G.M., & WILSON, E.G., 1959 - Bureau of Mineral Resources experimental water bore drilling, Canberra 1958. Bur. Min. Resour. Aust. Rec. 1959/53 (unpubl.).

NOAKES, L.C., 1954 - Geology and geomorphology of the Canberra Region. In H.L. WHITE (ed.) - CANBERRA, A NATION'S CAPITAL. Aust. Ass. Adv. Sci. Handb. for Canberra, 115-31.

OPIK, A.A., 1954 - Geology of the Canberra City District, in H.L. WHITE (ed.) - CANBERRA, A NATION'S CAPITAL. Aust. Ass. Adv. Sci. Handb. for Canberra, 131-48.

OPIK, A.A., 1958 - Geology of the Canberra City district. Bur. Min. Resour. Aust. Bull. 32.

WILSON, E.G., 1960 - Pumping tests and equipment in the A.C.T. Bur. Min. Resour. Aust. Rec. 1960/81 (unpubl.).

APPENDIX I

Example of Bore Advice Report

BUREAU OF MINERAL RESOURCES

GROUNDWATER SURVEY No. 1967/3

LOCATION: Parish Kowen Section 92 Block 3

Name of property:	'Kowenlea'
Owner:	Mr H.J. Harris
Postal address:	R.M.B. 252, Sutton Rd, Kowen, A.C.T.

OWNER'S REQUIREMENTS

Water required for:	Stock (including stud animals), garden and septic tank
Quantity required:	Average 400 g.p.d.; summer maximum 3000 g.p.d.
Quality required:	Preferably 1000 ppm
Location preferred:	South of homestead; reticulation to One Tree Hill
Probable type of Pump:	Electro-submersible
Location of electric supply:	Single-phase from transformer at south side of homestead

HYDROGEOLOGICAL REPORT

Geology: Lithology and age: Interbedded Middle Ordovician greywackes and slates. Greywacke dominant in lower section of sequence

Strike: N.20E Dip: Axis of anticline near quartz reef. Eastern limb dips 70° to east. Western limb 80° to west.

Faulting, jointing, lineaments: Quartz-filled fault striking N10°E and dipping 70° to E. Greywackes are well jointed at crest of anticline near fault

Geomorphology: Gentle east-west gully on rolling country with thin soil cover overlying deeply weathered sediments. Slates more deeply weathered than grey-wacke. Vegetation - natural savannah.

<u>Hydrology</u>: Discharge: Discharge of groundwater impeded by belt of deeply weathered thick sequence of slate cutting gully SE of homestead.

Recharge: Poor except where fault zone cuts gully and where fractured greywackes crop out.

Potentiometric surface: 10-15 feet below surface. Maximum variation likely over 10 years: 10 feet below and 5 feet above present level.

BORE SITE:

General location: 200 yards SSE of homestead

Sheet: 1:250 000 Canberra; 1-mile Canberra; Grid ref. 305425

Forecast:* Probable depth: 100 ft ± 15 ft. Owner should seek geologist's advice before drilling deeper than 150 ft if no water is encountered by that depth.

Probable logs: Soil 0-5 ft
Weathered slate 5-20 ft
Slate 20-60 ft
Slate & greywacke 60-90 ft
Quartz reef +90 ft

Probable depth of aquifers: Main supply about 90 ft

Standing water level: About 15 ft

Yield: 1500 gph ± 500 gph

Quality: About 800 ppm

Possible pollution: Negligible if drain from septic tank is removed.

Reason for location: To intersect quartz-filled fault which has good joint permeability and good recharge. Flat-lying beds on axis will not deflect drill.

*(This is a geological assessment of results expected and is for the general guidance only of interested parties. It is not a guarantee of drilling conditions and the final performance of the bore).

Recommendations for testing and development:	Yield:	At least 90 minute bailing test. Fault zone may carry considerable clay. Bail until water loses milkiness and development is complete.
	Quality:	Collect sample for chemical analysis when water becomes clear during test. Check analysis, particularly for sulphate.

General comments:

If sulphate is low and more water is required in years to come; it will be better to deepen this hole than to sink a second bore as yield of this hole will increase down to 200 ft because of fault zone.

Geologist .... Date 3/2/67

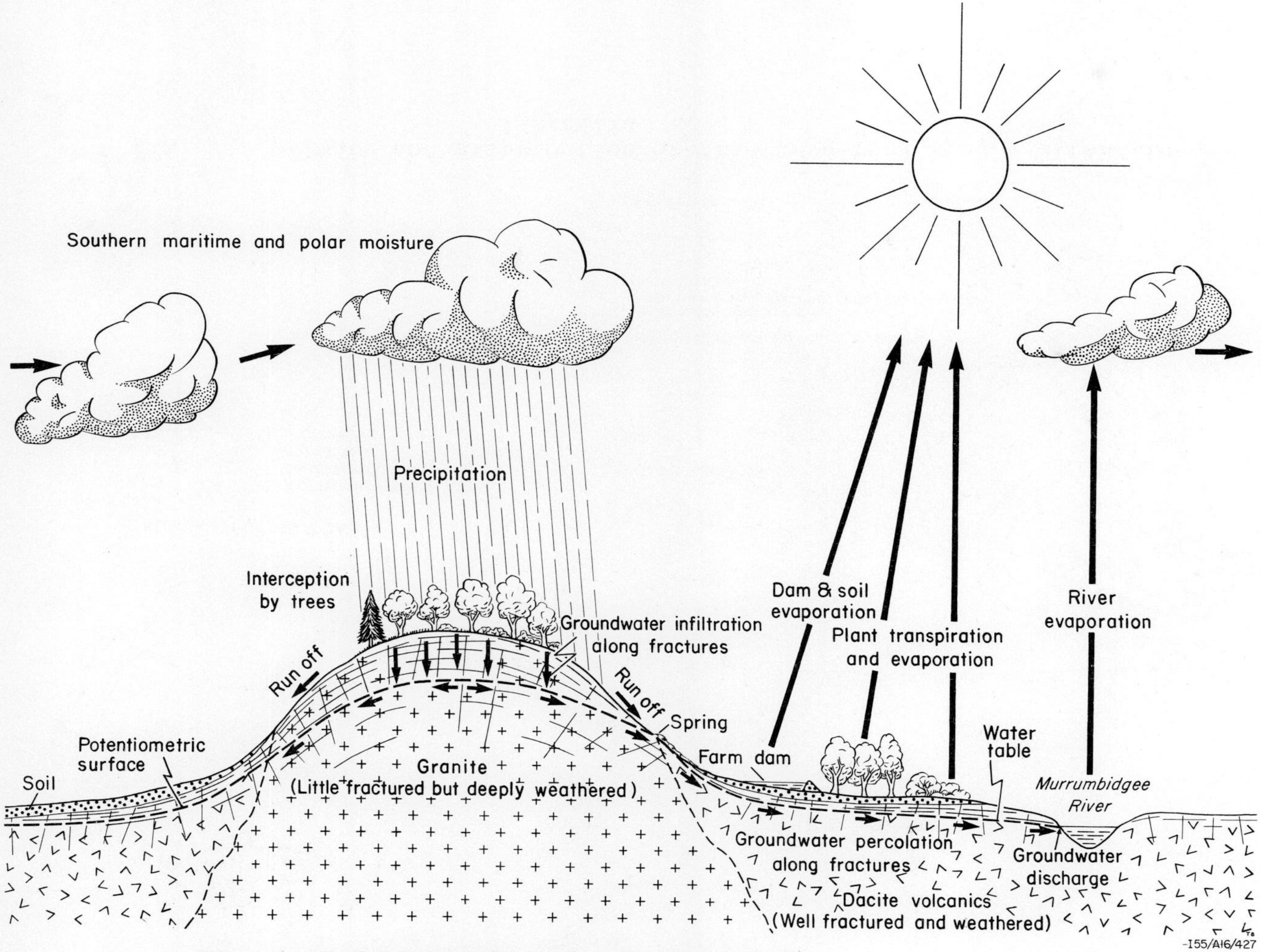

Fig. 1 Hydrological cycle in the A.C.T.

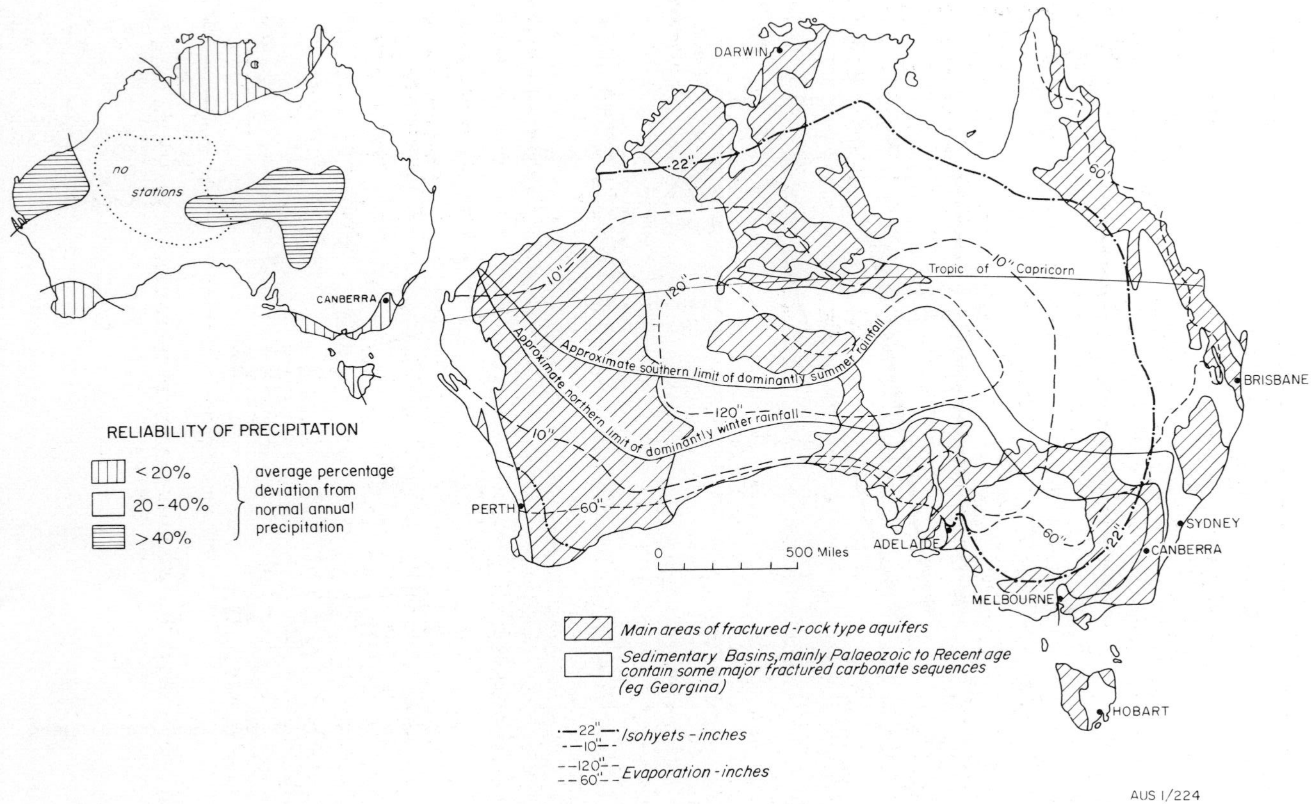

Fig. 2. Climate and distribution of fractured-rock type aquifers in Australia.

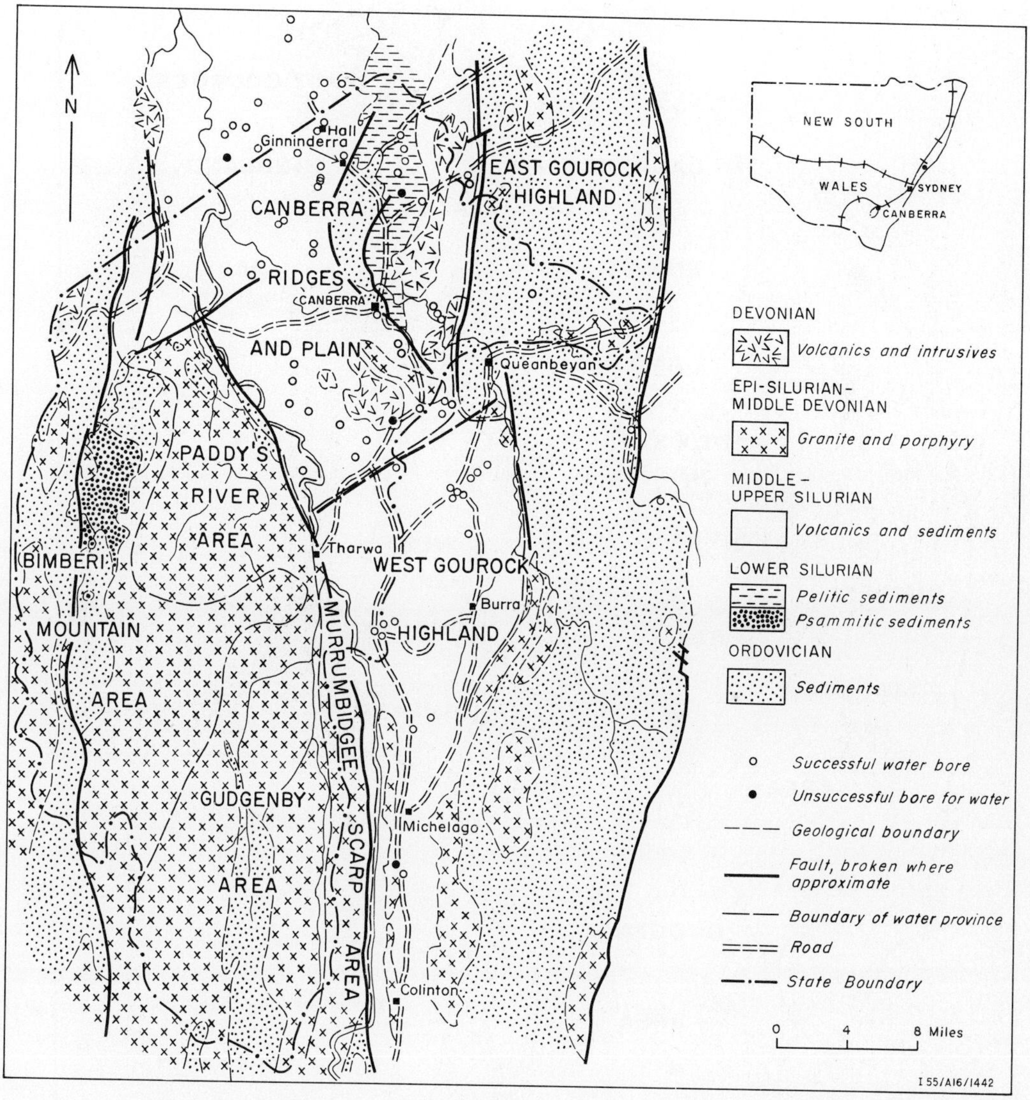

Fig. 3. A.C.T. groundwater provinces - geology and bores.

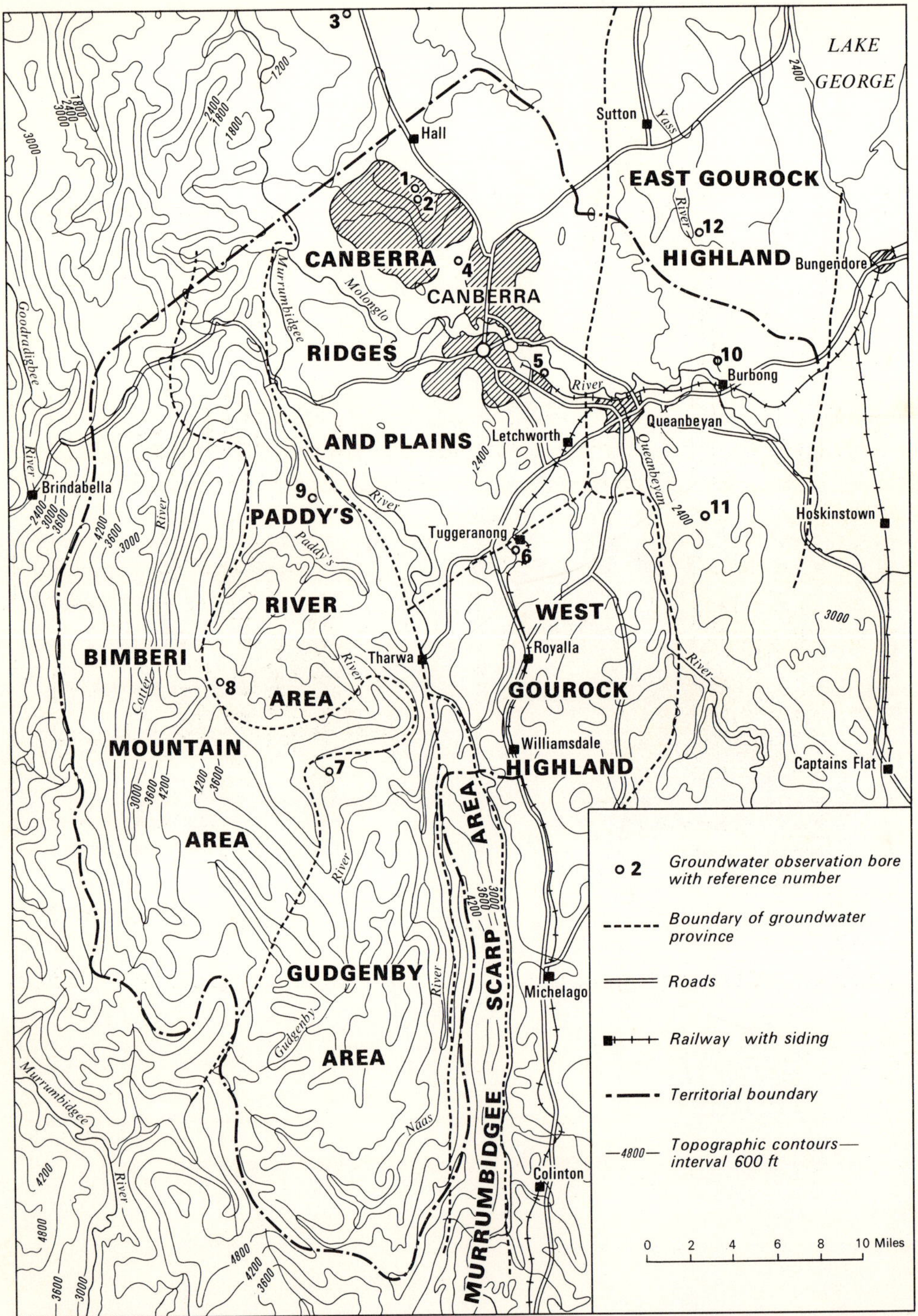

Fig. 4. Groundwater provinces of the A.C.T. (showing topography and location of observation bores).

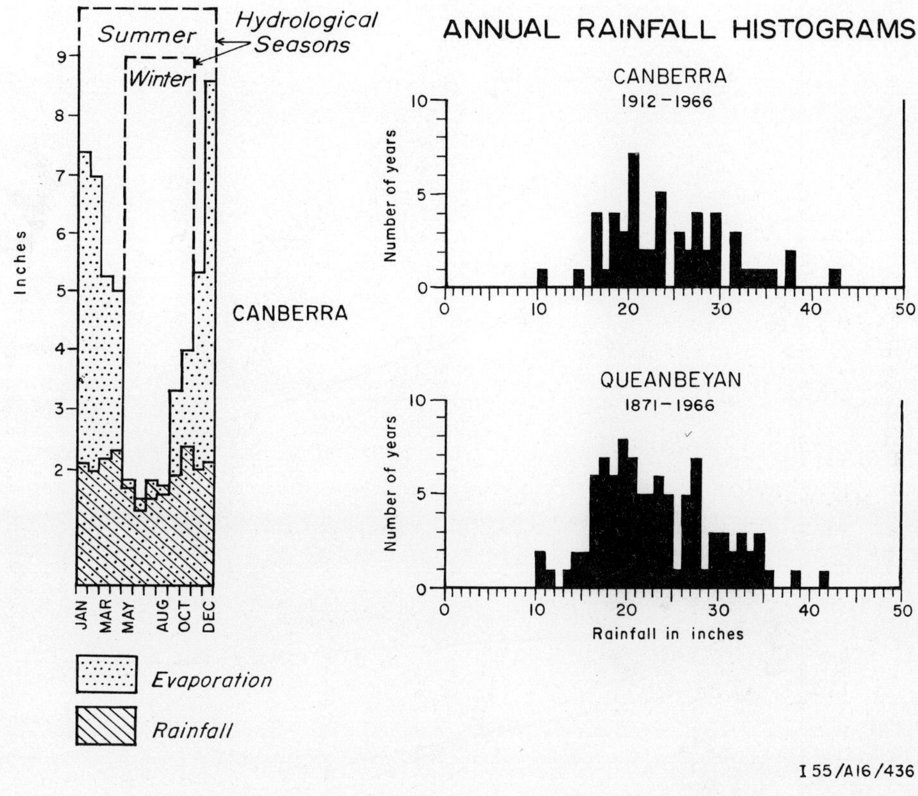

Fig. 5. Climatological data in the Canberra-Queanbeyan area.

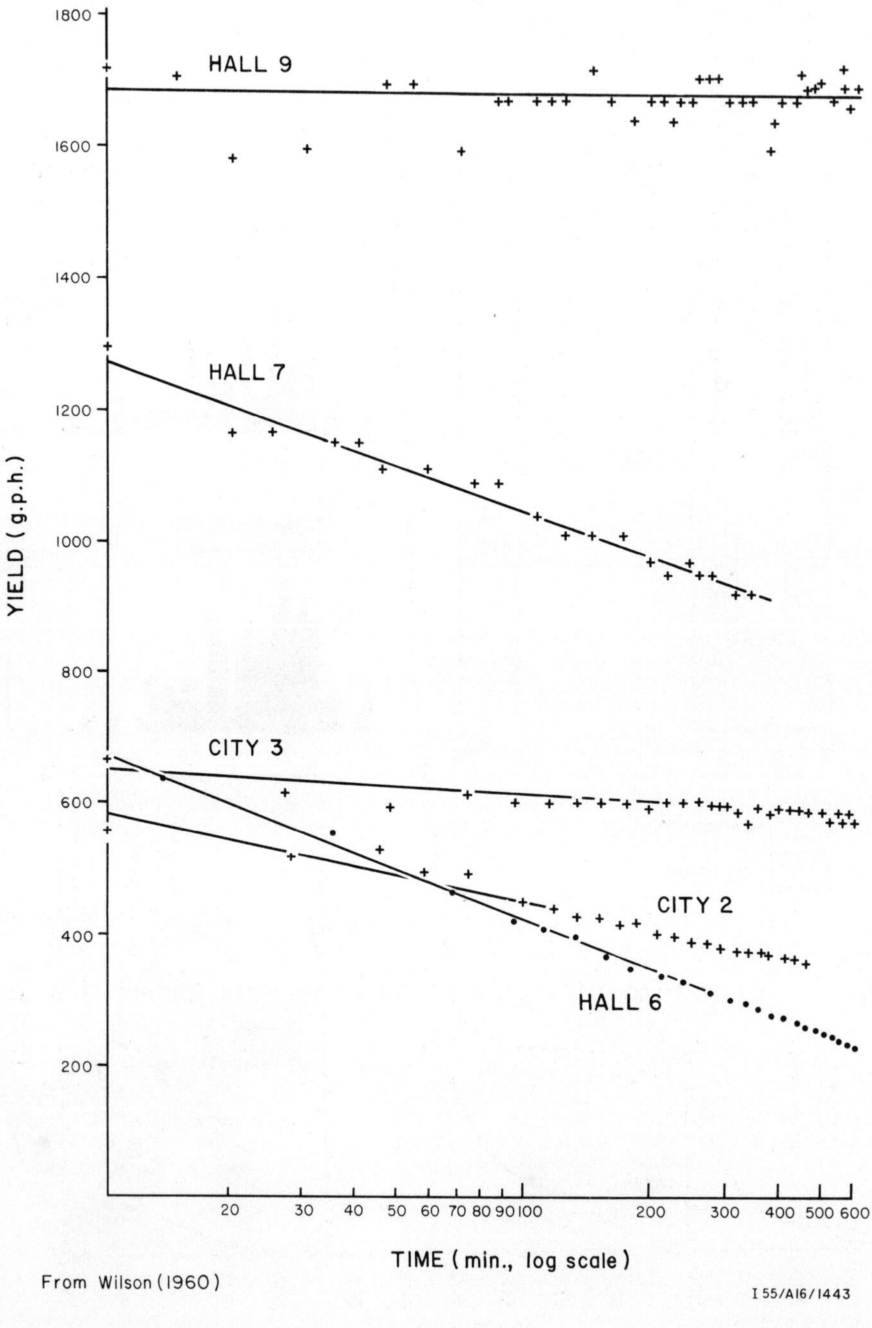

Fig. 6. A.C.T. and environs - pumping tests.

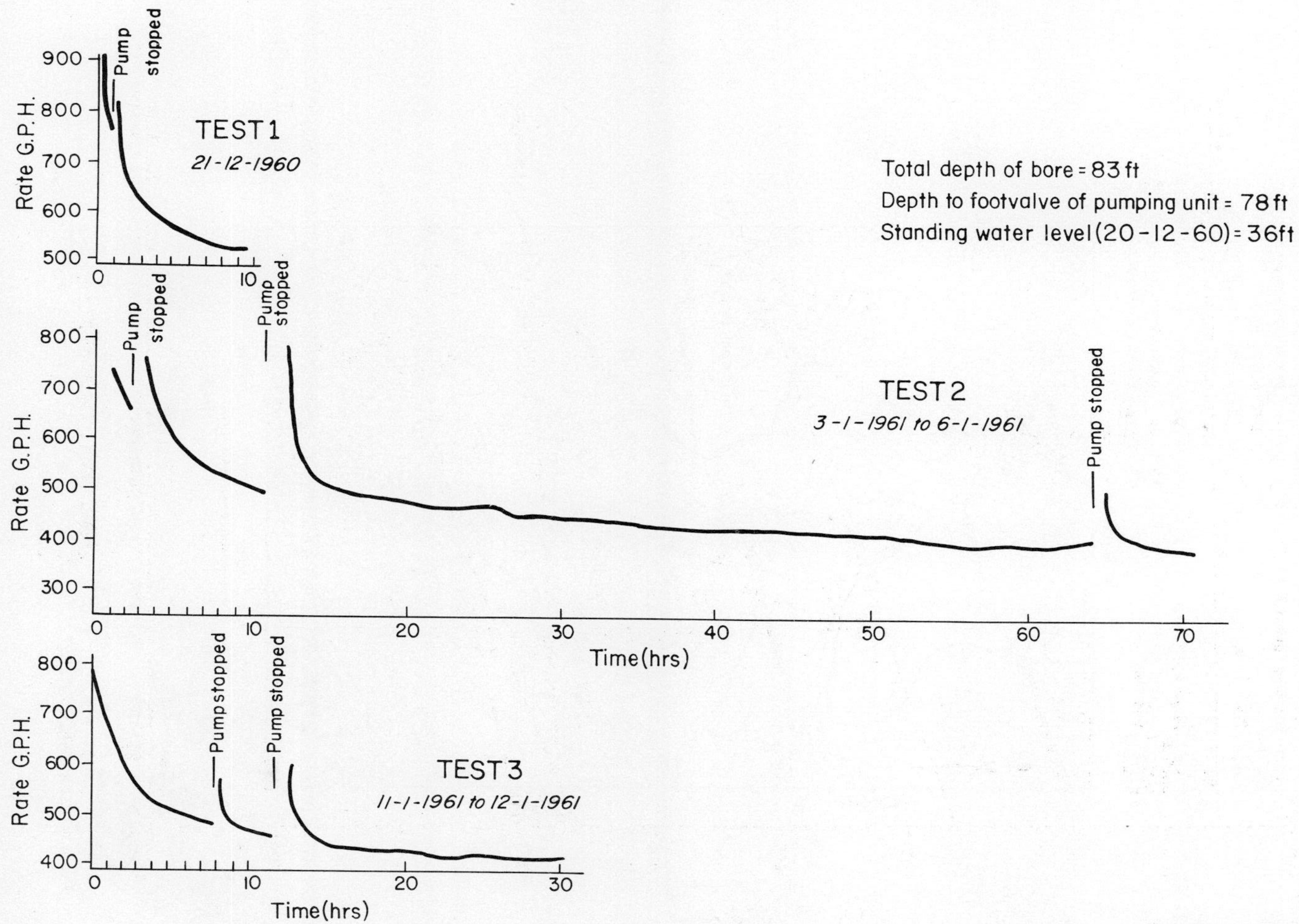

I 55/A16/440

Fig. 7. Pumping Test - Hall school bore, A.C.T.

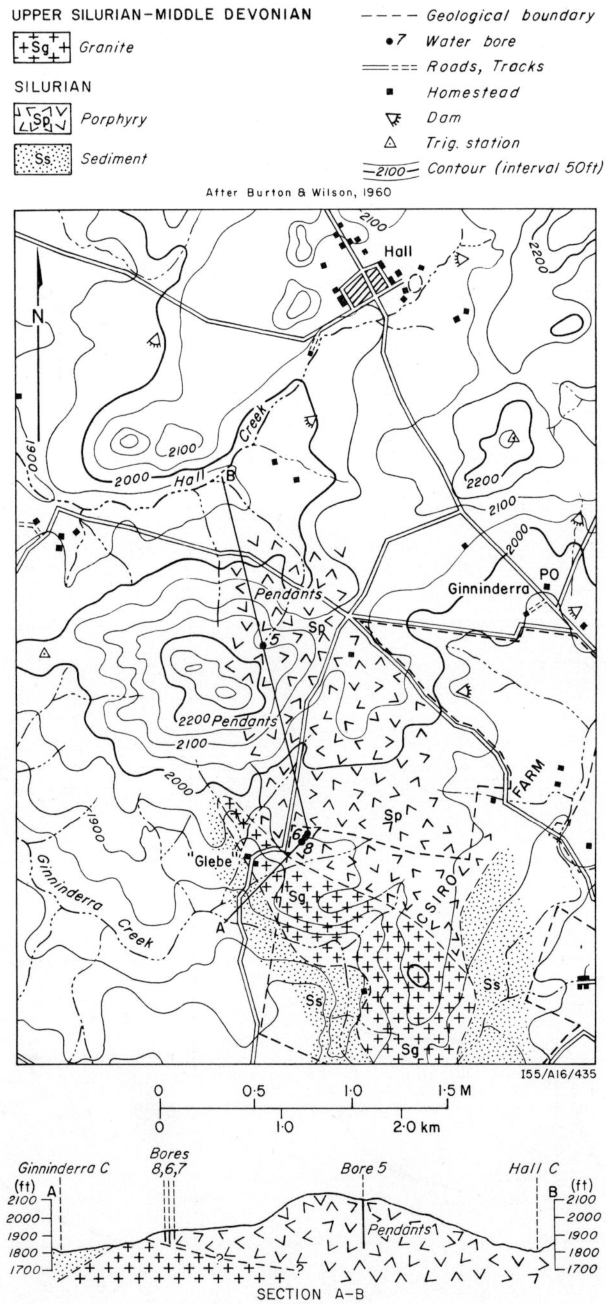

Fig. 8. Experimental water bores - Belconnen, A.C.T.

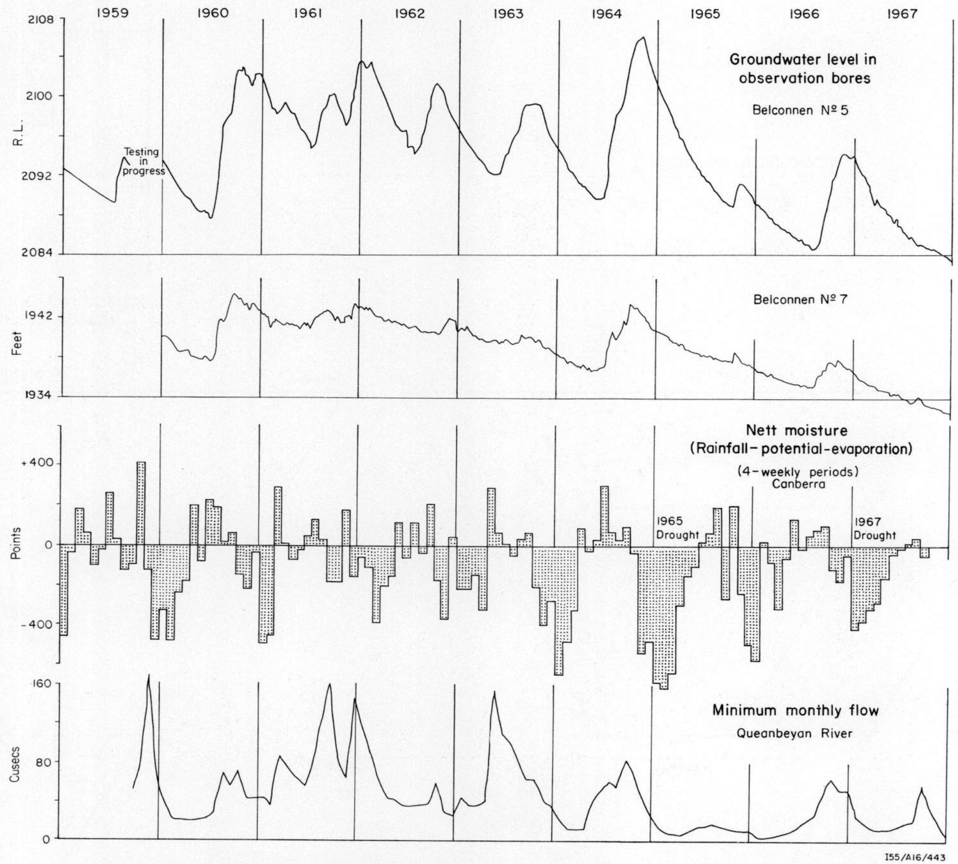

Fig. 9. Hydrographic data - A.C.T. and environs.

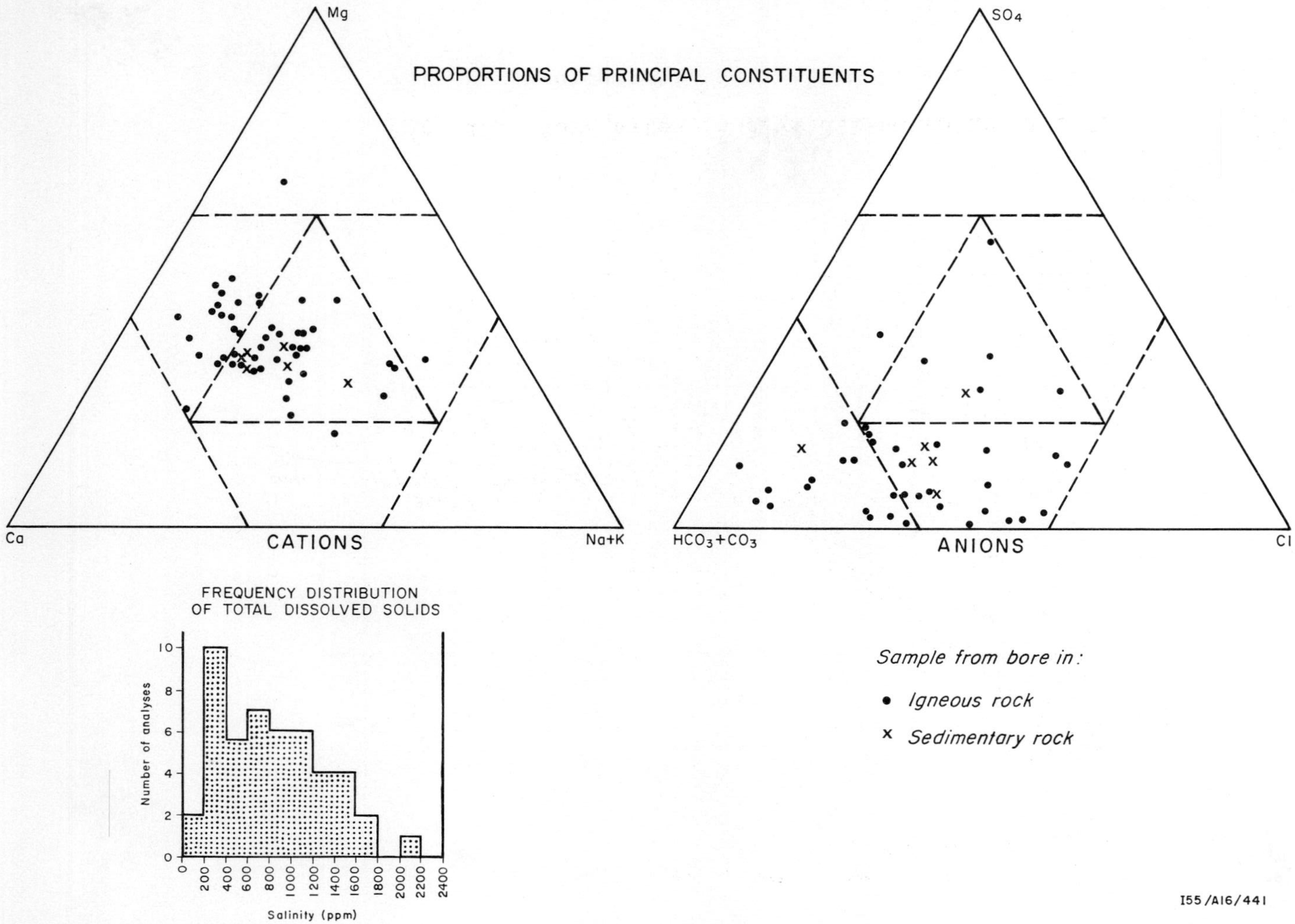

Fig. 10. Chemical composition of groundwater from water bores - A.C.T. and environs.

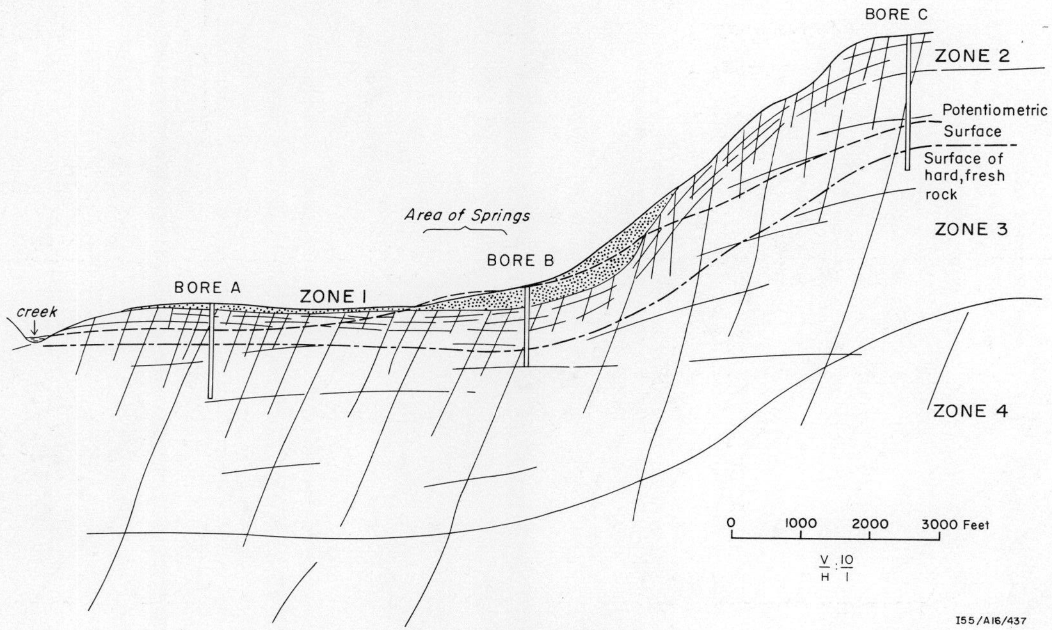

Fig. 11. Bore sites in crystalline rocks, A.C.T.

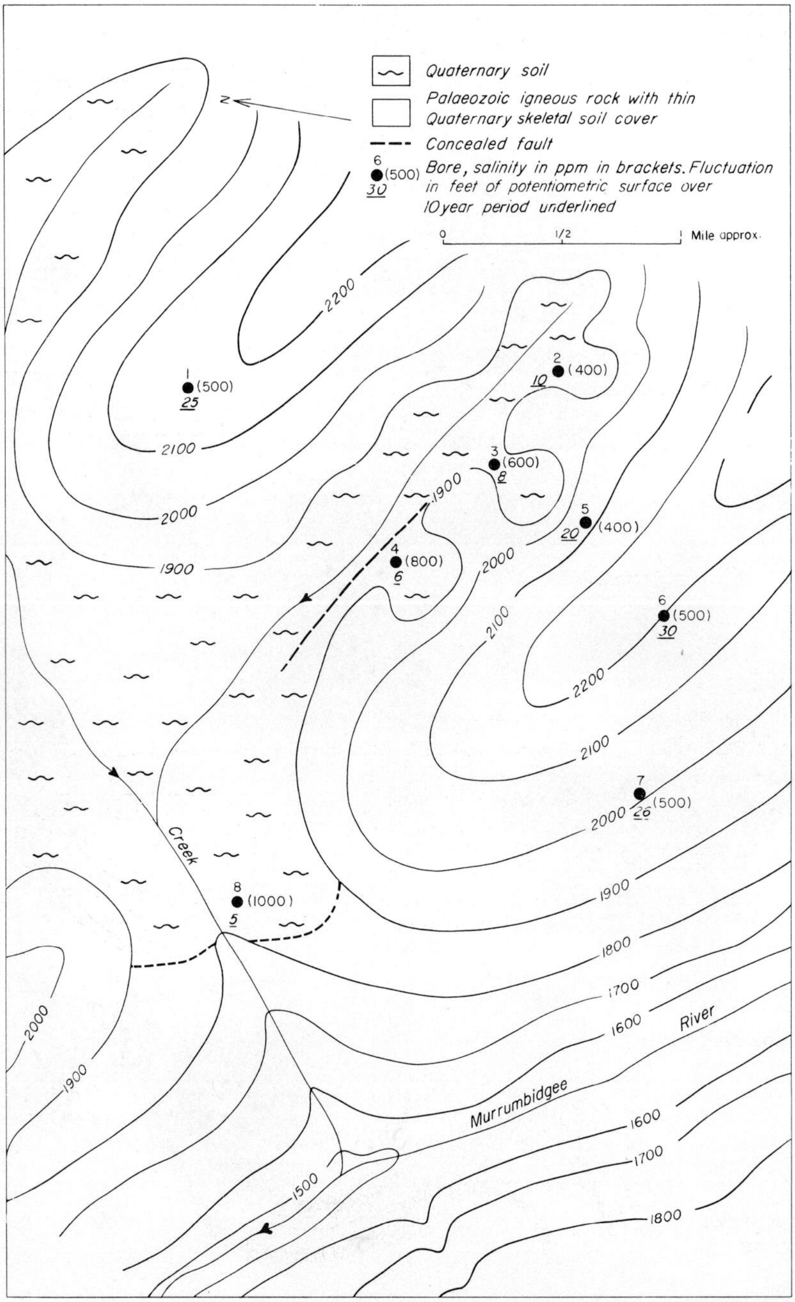

I55/A16/1444

Fig. 12. Geomorphological aspects of bore sites in the A.C.T.

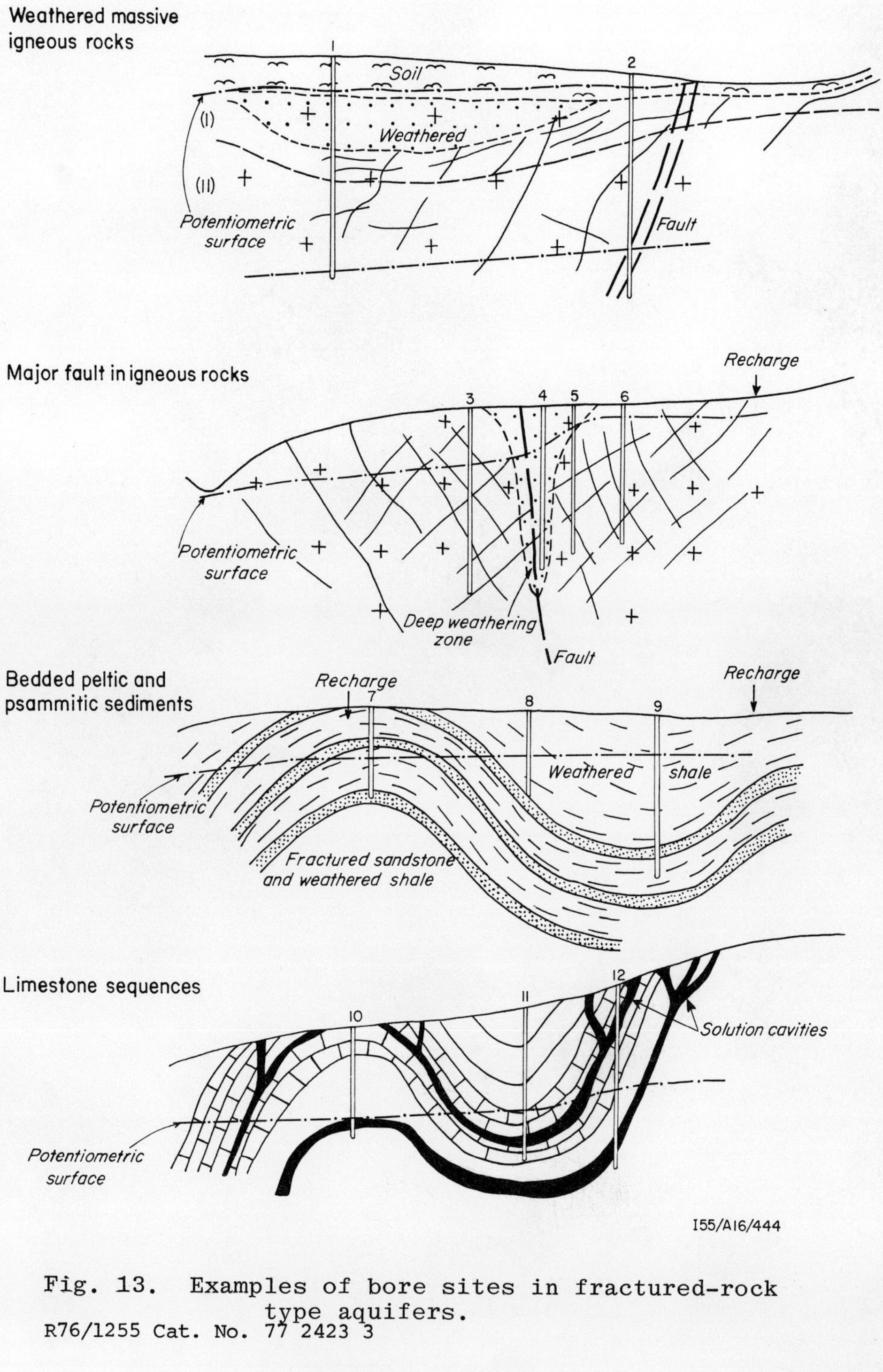

Fig. 13. Examples of bore sites in fractured-rock type aquifers.

R76/1255 Cat. No. 77 2423 3